EVOLUTION
& NATURAL SELECTION

EVOLUTION
& NATURAL SELECTION

August Weismann on
Darwin and Lamarck

Excerpted from Darwin and Modern Science by A.C. Seward & Others

Slightly edited by

David Christopher Lane

MSAC Philosophy Group
Walnut, California

EVOLUTION & NATURAL SELECTION

Format Copyright © 2018 by MSAC Philosophy Group

Second Edition| Trade paperback

ISBN: 978-1-56543-206-2

General Editors: Dr. Andrea Diem and Dr. David Lane

MSAC *Philosophy Group* was founded in 1990 and is designed to provide a wide range of materials (from books to magazines to films to audio presentations to interactive texts) on such subjects as evolutionary biology, quantum theory, neuroscience, and critical studies in religion and philosophy. All books are sold on a not for profit basis. Free PDF versions are made available whenever possible. In addition, there is a growing collection of audio books specifically created for students at Mt. San Antonio College and the community at large. A large series of original movies have also been produced which touch on such topics as artificial intelligence, eliminative materialism, consciousness, and skepticism.

Typeset: Main body, 14 pt. Garamond | Published in the USA

Dedication

To a proper understanding of Darwin

I. THE IDEA OF SELECTION

Many and diverse were the discoveries made by Charles Darwin in the course of a long and strenuous life, but none of them has had so far-reaching an influence on the science and thought of his time as the theory of selection. I do not believe that the theory of evolution would have made its way so easily and so quickly after Darwin took up the cudgels in favour of it, if he had not been able to support it by a principle which was capable of solving, in a simple manner, the greatest riddle that living nature presents to us,—I mean the purposiveness of every living form relative to the conditions of its life and its marvellously exact adaptation to these.

Everyone knows that Darwin was not alone in discovering the principle of selection, and that the same idea occurred simultaneously and independently to Alfred Russel Wallace. At the memorable meeting of the Linnean Society on 1st July, 1858, two papers were read (communicated by Lyell and Hooker) both setting forth the same idea of selection. One was written by Charles Darwin in Kent, the other by Alfred Wallace in Ternate, in the Malay Archipelago. It was a splendid proof of the magnanimity of these two investigators, that they thus, in all friendliness and without envy, united in laying their ideas before a scientific tribunal: their names will always shine side by side as two of the brightest stars in the scientific sky.

But it is with Charles Darwin that I am here chiefly concerned, since this paper is intended to aid in the commemoration of the hundredth anniversary of his birth.

The idea of selection set forth by the two naturalists was at the time absolutely new, but it was also so simple that Huxley could say of it later, "How extremely stupid not to have thought of that." As Darwin was led to the general doctrine of descent, not through the labours of his predecessors in the early years of the century, but by his own observations, so it was in regard to the principle of selection. He was struck by the innumerable cases of adaptation, as, for instance, that of the woodpeckers and tree-frogs to climbing, or the hooks and feather-like appendages of seeds, which aid in the distribution of plants, and he said to himself that an explanation of adaptations was the first thing to be sought for in attempting to formulate a theory of evolution.

But since adaptations point to changes which have been undergone by the ancestral forms of existing species, it is necessary, first of all, to inquire how far species in general are variable. Thus Darwin's attention was directed in the first place to the phenomenon of variability, and the use man has made of this, from very early times, in the breeding of his domesticated animals and cultivated plants. He inquired carefully how breeders set to work, when they wished to modify the structure and appearance of a species to their own ends, and it was soon clear to him that selection for breeding purposes played the chief part.

But how was it possible that such processes should occur in free nature? Who is here the breeder, making the selection, choosing out one individual to bring forth offspring and rejecting others? That was the problem that for a long time remained a riddle to him.

Darwin himself relates how illumination suddenly came to him. He had been reading, for his own pleasure, Malthus' book on Population, and, as he had long known from numerous observations, that every species gives rise to many more descendants than ever attain to maturity, and that, therefore, the greater number of the descendants of a species perish without reproducing, the idea came to him that the decision as to which member of a species was to perish, and which was to attain to maturity and reproduction might not be a matter of chance, but might be determined by the constitution of the individuals themselves, according as they were more or less fitted for survival. With this idea the foundation of the theory of selection was laid.

In artificial selection the breeder chooses out for pairing only such individuals as possess the character desired by him in a somewhat higher degree than the rest of the race. Some of the descendants inherit this character, often in a still higher degree, and if this method be pursued throughout several generations, the race is transformed in respect of that particular character.

Natural selection depends on the same three factors as artificial selection: on variability, inheritance, and selection for breeding, but this last is here carried out not by a breeder but by what Darwin called the "struggle for existence." This last factor is one of the special features of the Darwinian conception of nature. That there are carnivorous animals which take heavy toll in every generation of the progeny of the animals on which they prey, and that there are herbivores which decimate the plants in every generation had long been known, but it is only since Darwin's time that sufficient attention has been paid to the facts that, in addition to this regular destruction, there exists between the members of a species a keen competition for space and food, which limits multiplication, and that numerous individuals of each species perish because of unfavourable climatic conditions. The "struggle for existence," which Darwin regarded as taking the place of the human breeder in free nature, is not a direct struggle between carnivores and their prey, but is the assumed competition for survival between individuals of the same species, of which, on an average, only those survive to reproduce which have the greatest power of resistance, while the others, less favourably constituted, perish early. This struggle is so keen, that, within a limited area, where the conditions of life have long remained unchanged, of every species, whatever be the degree of fertility, only two, on an average, of the descendants of each pair survive; the others succumb either to enemies, or to disadvantages of climate, or to accident. A high degree of fertility is thus not an indication of the special success of a species, but of the numerous dangers that have attended its evolution. Of the six young brought forth by a pair of elephants in the course of their lives only two survive in a given area; similarly, of the millions of eggs which two thread-worms leave behind them only two survive. It is thus possible to estimate the dangers which threaten a species by its ratio of elimination, or, since this cannot be done directly, by its fertility.

Although a great number of the descendants of each generation fall victims to accident, among those that remain it is still the greater or lesser fitness of the organism that determines the "selection for breeding purposes," and it would be incomprehensible if, in this competition, it were not ultimately, that is, on an average, the best equipped which survive, in the sense of living long enough to reproduce.

Thus the principle of natural selection is the selection of the best for reproduction, whether the "best" refers to the whole constitution, to one or more parts of the organism, or to one or more stages of development. Every organ, every part, every character of an animal, fertility and intelligence included, must be improved in this manner, and be gradually brought up in the course of generations to its highest attainable state of perfection. And not only may improvement of parts be brought about in this way, but new parts and organs may arise, since, through the slow and minute steps of individual or "fluctuating" variations, a part may be added here or dropped out there, and thus something new is produced.

The principle of selection solved the riddle as to how what was purposive could conceivably be brought about without the intervention of a directing power, the riddle which animate nature presents to our intelligence at every turn, and in face of which the mind of a Kant could find no way out, for he regarded a solution of it as not to be hoped for. For, even if we were to assume an evolutionary force that is continually transforming the most primitive and the simplest forms of life into ever higher forms, and the homogeneity of primitive times into the infinite variety of the present, we should still be unable to infer from this alone how each of the numberless forms adapted to particular conditions of life should have appeared precisely at the right moment in the history of the earth to which their adaptations were appropriate, and precisely at the proper place in which all the conditions of life to which they were adapted occurred: the humming-birds at the same time as the flowers; the trichina at the same time as the pig; the bark-coloured moth at the same time as the oak, and the wasp-like moth at the same time as the wasp which protects it. Without processes of selection we should be obliged to assume a "pre-established harmony" after the famous Leibnitzian model, by means of which the clock of the evolution of organisms is so regulated as to strike in exact synchronism with that of the history of the earth! All forms of life are strictly adapted to the conditions of their life, and can persist under these conditions alone.

There must therefore be an intrinsic connection between the conditions and the structural adaptations of the organism, and, since the conditions of life cannot be determined by the animal itself, the adaptations must be called forth by the conditions.

The selection theory teaches us how this is conceivable, since it enables us to understand that there is a continual production of what is non-purposive as well as of what is purposive, but the purposive alone survives, while the non-purposive perishes in the very act of arising. This is the old wisdom taught long ago by Empedocles.

II. THE LAMARCKIAN PRINCIPLE

Lamarck, as is well known, formulated a definite theory of evolution at the beginning of the nineteenth century, exactly fifty years before the Darwin-Wallace principle of selection was given to the world. This brilliant investigator also endeavoured to support his theory by demonstrating forces which might have brought about the transformations of the organic world in the course of the ages. In addition to other factors, he laid special emphasis on the increased or diminished use of the parts of the body, assuming that the strengthening or weakening which takes place from this cause during the individual life, could be handed on to the offspring, and thus intensified and raised to the rank of a specific character. Darwin also regarded this LAMARCKIAN PRINCIPLE, as it is now generally called, as a factor in evolution, but he was not fully convinced of the transmissibility of acquired characters.

As I have here to deal only with the theory of selection, I need not discuss the Lamarckian hypothesis, but I must express my opinion that there is room for much doubt as to the cooperation of this principle in evolution. Not only is it difficult to imagine how the transmission of functional modifications could take place, but, up to the present time, notwithstanding the endeavours of many excellent investigators, not a single actual proof of such inheritance has been brought forward. Semon's experiments on plants are, according to the botanist Pfeffer, not to be relied on, and even the recent, beautiful experiments made by Dr Kammerer on salamanders, cannot, as I hope to show elsewhere, be regarded as proof, if only because they do not deal at all with functional modifications, that is, with modifications brought about by use, and it is to these ALONE that the Lamarckian principle refers.

III. OBJECTIONS TO THE THEORY OF SELECTION

(a) Saltatory Evolution.

The Darwinian doctrine of evolution depends essentially on THE CUMULATIVE AUGMENTATION of minute variations in the direction of utility. But can such minute variations, which are undoubtedly continually appearing among the individuals of the same species, possess any selection-value; can they determine which individuals are to survive, and which are to succumb; can they be increased by natural selection till they attain to the highest development of a purposive variation?

To many this seems so improbable that they have urged a theory of evolution by leaps from species to species. Kolliker, in 1872, compared the evolution of species with the processes which we can observe in the individual life in cases of alternation of generations. But a polyp only gives rise to a medusa because it has itself arisen from one, and there can be no question of a medusa ever having arisen suddenly and de novo from a polyp-bud, if only because both forms are adapted in their structure as a whole, and in every detail to the conditions of their life. A sudden origin, in a natural way, of numerous adaptations is inconceivable. Even the degeneration of a medusoid from a free-swimming animal to a mere brood-sac (gonophore) is not sudden and saltatory, but occurs by imperceptible modifications throughout hundreds of years, as we can learn from the numerous stages of the process of degeneration persisting at the same time in different species.

If, then, the degeneration to a simple brood-sac takes place only by very slow transitions, each stage of which may last for centuries, how could the much more complex ascending evolution possibly have taken place by sudden leaps? I regard this argument as capable of further extension, for wherever in nature we come upon degeneration, it is taking place by minute steps and with a slowness that makes it not directly perceptible, and I believe that this in itself justifies us in concluding that the same must be true of ascending evolution. But in the latter case the goal can seldom be distinctly recognised while in cases of degeneration the starting-point of the process can often be inferred, because several nearly related species may represent different stages.

In recent years Bateson in particular has championed the idea of saltatory, or so-called discontinuous evolution, and has collected a number of cases in which more or less marked variations have suddenly appeared. These are taken for the most part from among domesticated animals which have been bred and crossed for a long time, and it is hardly to be wondered at that their much mixed and much influenced germ-plasm should, under certain conditions, give rise to remarkable phenomena, often indeed producing forms which are strongly suggestive of monstrosities, and which would undoubtedly not survive in free nature, unprotected by man. I should regard such cases as due to an intensified germinal selection—though this is to anticipate a little—and from this point of view it cannot be denied that they have a special interest. But they seem to me to have no significance as far as the transformation of species is concerned, if only because of the extreme rarity of their occurrence.

There are, however, many variations which have appeared in a sudden and saltatory manner, and some of these Darwin pointed out and discussed in detail: the copper beech, the weeping trees, the oak with "fern-like leaves," certain garden-flowers, etc. But none of them have persisted in free nature, or evolved into permanent types.

On the other hand, wherever enduring types have arisen, we find traces of a gradual origin by successive stages, even if, at first sight, their origin may appear to have been sudden. This is the case with seasonal dimorphism, the first known cases of which exhibited marked differences between the two generations, the winter and the summer brood. Take for instance the much discussed and studied form Vanessa (Araschnia) levana-prorsa. Here the differences between the two forms are so great and so apparently disconnected, that one might almost believe it to be a sudden mutation, were it not that old transition-stages can be called forth by particular temperatures, and we know other butterflies, as for instance our Garden Whites, in which the differences between the two generations are not nearly so marked; indeed, they are so little apparent that they are scarcely likely to be noticed except by experts. Thus here again there are small initial steps, some of which, indeed, must be regarded as adaptations, such as the green-sprinkled or lightly tinted under-surface which gives them a deceptive resemblance to parsley or to Cardamine leaves.

Even if saltatory variations do occur, we cannot assume that these have ever led to forms which are capable of survival under the conditions of wild life. Experience has shown that in plants which have suddenly varied the power of persistence is diminished. Korschinksky attributes to them weaknesses of organisation in general; "they bloom late, ripen few of their seeds, and show great sensitiveness to cold." These are not the characters which make for success in the struggle for existence.

We must briefly refer here to the views—much discussed in the last decade—of H. de Vries, who believes that the roots of transformation must be sought for in saltatory variations arising from internal causes, and distinguishes such mutations, as he has called them, from ordinary individual variations, in that they breed true, that is, with strict inbreeding they are handed on pure to the next generation. I have elsewhere endeavoured to point out the weaknesses of this theory ("Vortrage uber Descendenztheorie", Jena, 1904, II. 269. English Translation London, 1904, II. page 317.), and I am the less inclined to return to it here that it now appears (See Poulton, "Essays on Evolution", Oxford, 1908, pages xix-xxii.) that the far-reaching conclusions drawn by de Vries from his observations on the Evening Primrose, Oenothera lamarckiana, rest upon a very insecure foundation. The plant from which de Vries saw numerous "species"—his "mutations"—arise was not, as he assumed, a wild species that had been introduced to Europe from America, but was probably a hybrid form which was first discovered in the Jardin des Plantes in Paris, and which does not appear to exist anywhere in America as a wild species.

This gives a severe shock to the "Mutation theory," for the other actually wild species with which de Vries experimented showed no "mutations" but yielded only negative results.

Thus we come to the conclusion that Darwin ("Origin of Species" (6th edition), pages 176 et seq.) was right in regarding transformations as taking place by minute steps, which, if useful, are augmented in the course of innumerable generations, because their possessors more frequently survive in the struggle for existence.

(b) Selection-value of the initial steps.

Is it possible that the significant deviations which we know as "individual variations" can form the beginning of a process of selection? Can they decide which is to perish and which to survive? To use a phrase of Romanes, can they have selection-value?

Darwin himself answered this question, and brought together many excellent examples to show that differences, apparently insignificant because very small, might be of decisive importance for the life of the possessor. But it is by no means enough to bring forward cases of this kind, for the question is not merely whether finished adaptations have selection-value, but whether the first beginnings of these, and whether the small, I might almost say minimal increments, which have led up from these beginnings to the perfect adaptation, have also had selection-value. To this question even one who, like myself, has been for many years a convinced adherent of the theory of selection, can only reply: we must assume so, but we cannot prove it in any case. It is not upon demonstrative evidence that we rely when we champion the doctrine of selection as a scientific truth; we base our argument on quite other grounds. Undoubtedly there are many apparently insignificant features, which can nevertheless be shown to be adaptations—for instance, the thickness of the basin-shaped shell of the limpets that live among the breakers on the shore. There can be no doubt that the thickness of these shells, combined with their flat form, protects the animals from the force of the waves breaking upon them,—but how have they become so thick? What proportion of thickness was sufficient to decide that of two variants of a limpet one should survive, the other be eliminated? We can say nothing more than that we infer from the present state of the shell, that it must have varied in regard to differences in shell-thickness, and that these differences must have had selection-value,—no proof therefore, but an assumption which we must show to be convincing.

For a long time the marvellously complex radiate and lattice-work skeletons of Radiolarians were regarded as a mere outflow of "Nature's infinite wealth of form," as an instance of a purely morphological character with no biological significance. But recent investigations have shown that these, too, have an adaptive significance (Hacker). The same thing has been shown by Schutt in regard to the lowly unicellular plants, the Peridineae, which abound alike on the surface of the ocean and in its depths. It has been shown that the long skeletal processes which grow out from these organisms have significance not merely as a supporting skeleton, but also as an extension of the superficial area, which increases the contact with the water-particles, and prevents the floating organisms from sinking. It has been established that the processes are considerably shorter in the colder layers of the ocean, and that they may be twelve times as long (Chun, "Reise der Valdivia", Leipzig, 1904.) in the warmer layers, thus corresponding to the greater or smaller amount of friction which takes place in the denser and less dense layers of the water.

The Peridineae of the warmer ocean layers have thus become long-rayed, those of the colder layers short-rayed, not through the direct effect of friction on the protoplasm, but through processes of selection, which favoured the longer rays in warm water, since they kept the organism afloat, while those with short rays sank and were eliminated. If we put the question as to selection-value in this case, and ask how great the variations in the length of processes must be in order to possess selection-value; what can we answer except that these variations must have been minimal, and yet sufficient to prevent too rapid sinking and consequent elimination? Yet this very case would give the ideal opportunity for a mathematical calculation of the minimal selection-value, although of course it is not feasible from lack of data to carry out the actual calculation.

But even in organisms of more than microscopic size there must frequently be minute, even microscopic differences which set going the process of selection, and regulate its progress to the highest possible perfection.

Many tropical trees possess thick, leathery leaves, as a protection against the force of the tropical rain drops. The direct influence of the rain cannot be the cause of this power of resistance, for the leaves, while they were still thin, would simply have been torn to pieces. Their toughness must therefore be referred to selection, which would favour the trees with slightly thicker leaves, though we cannot calculate with any exactness how great the first stages of increase in thickness must have been. Our hypothesis receives further support from the fact that, in many such trees, the leaves are drawn out into a beak-like prolongation (Stahl and Haberlandt) which facilitates the rapid falling off of the rain water, and also from the fact that the leaves, while they are still young, hang limply down in bunches which offer the least possible resistance to the rain. Thus there are here three adaptations which can only be interpreted as due to selection. The initial stages of these adaptations must undoubtedly have had selection-value.

But even in regard to this case we are reasoning in a circle, not giving "proofs," and no one who does not wish to believe in the selection-value of the initial stages can be forced to do so. Among the many pieces of presumptive evidence a particularly weighty one seems to me to be the smallness of the steps of progress which we can observe in certain cases, as for instance in leaf-imitation among butterflies, and in mimicry generally. The resemblance to a leaf, for instance of a particular Kallima, seems to us so close as to be deceptive, and yet we find in another individual, or it may be in many others, a spot added which increases the resemblance, and which could not have become fixed unless the increased deceptiveness so produced had frequently led to the overlooking of its much persecuted possessor. But if we take the selection-value of the initial stages for granted, we are confronted with the further question which I myself formulated many years ago: How does it happen that the necessary beginnings of a useful variation are always present? How could insects which live upon or among green leaves become all green, while those that live on bark become brown? How have the desert animals become yellow and the Arctic animals white? Why were the necessary variations always present? How could the green locust lay brown eggs, or the privet caterpillar develop white and lilac-coloured lines on its green skin?

It is of no use answering to this that the question is wrongly formulated (Plate, "Selektionsprinzip u. Probleme der Artbildung" (3rd edition), Leipzig, 1908.) and that it is the converse that is true; that the process of selection takes place in accordance with the variations that present themselves. This proposition is undeniably true, but so also is another, which apparently negatives it: the variation required has in the majority of cases actually presented itself. Selection cannot solve this contradiction; it does not call forth the useful variation, but simply works upon it. The ultimate reason why one and the same insect should occur in green and in brown, as often happens in caterpillars and locusts, lies in the fact that variations towards brown presented themselves, and so also did variations towards green: the kernel of the riddle lies in the varying, and for the present we can only say, that small variations in different directions present themselves in every species. Otherwise so many different kinds of variations could not have arisen. I have endeavoured to explain this remarkable fact by means of the intimate processes that must take place within the germ-plasm, and I shall return to the problem when dealing with "germinal selection."

We have, however, to make still greater demands on variation, for it is not enough that the necessary variation should occur in isolated individuals, because in that case there would be small prospect of its being preserved, notwithstanding its utility. Darwin at first believed, that even single variations might lead to transformation of the species, but later he became convinced that this was impossible, at least without the cooperation of other factors, such as isolation and sexual selection.

In the case of the green caterpillars with bright longitudinal stripes, numerous individuals exhibiting this useful variation must have been produced to start with. In all higher, that is, multicellular organisms, the germ-substance is the source of all transmissible variations, and this germ-plasm is not a simple substance but is made up of many primary constituents. The question can therefore be more precisely stated thus: How does it come about that in so many cases the useful variations present themselves in numbers just where they are required, the white oblique lines in the leaf-caterpillar on the under surface of the body, the accompanying coloured stripes just above them? And, further, how has it come about that in grass caterpillars, not oblique but longitudinal stripes, which are more effective for concealment among grass and plants, have been evolved? And finally, how is it that the same Hawk-moth caterpillars, which to-day show oblique stripes, possessed longitudinal stripes in Tertiary times? We can read this fact from the history of their development, and I have before attempted to show the biological significance of this change of colour. ("Studien zur Descendenz-Theorie" II., "Die Enstehung der Zeichnung bei den Schmetterlings-raupen," Leipzig, 1876.)

For the present I need only draw the conclusion that one and the same caterpillar may exhibit the initial stages of both, and that it depends on the manner in which these marking elements are intensified and combined by natural selection whether whitish longitudinal or oblique stripes should result. In this case then the "useful variations" were actually "always there," and we see that in the same group of Lepidoptera, e.g. species of Sphingidae, evolution has occurred in both directions according to whether the form lived among grass or on broad leaves with oblique lateral veins, and we can observe even now that the species with oblique stripes have longitudinal stripes when young, that is to say, while the stripes have no biological significance. The white places in the skin which gave rise, probably first as small spots, to this protective marking could be combined in one way or another according to the requirements of the species. They must therefore either have possessed selection-value from the first, or, if this was not the case at their earliest occurrence, there must have been some other factors which raised them to the point of selection-value. I shall return to this in discussing germinal selection. But the case may be followed still farther, and leads us to the same alternative on a still more secure basis.

Many years ago I observed in caterpillars of Smerinthus populi (the poplar hawk-moth), which also possess white oblique stripes, that certain individuals showed red spots above these stripes; these spots occurred only on certain segments, and never flowed together to form continuous stripes. In another species (Smerinthus tiliae) similar blood-red spots unite to form a line-like coloured seam in the last stage of larval life, while in S. ocellata rust-red spots appear in individual caterpillars, but more rarely than in S. Populi, and they show no tendency to flow together.

Thus we have here the origin of a new character, arising from small beginnings, at least in S. tiliae, in which species the coloured stripes are a normal specific character. In the other species, S. populi and S. ocellata, we find the beginnings of the same variation, in one more rarely than in the other, and we can imagine that, in the course of time, in these two species, coloured lines over the oblique stripes will arise. In any case these spots are the elements of variation, out of which coloured lines MAY be evolved, if they are combined in this direction through the agency of natural selection. In S. populi the spots are often small, but sometimes it seems as though several had united to form large spots. Whether a process of selection in this direction will arise in S. populi and S. ocellata, or whether it is now going on cannot be determined, since we cannot tell in advance what biological value the marking might have for these two species. It is conceivable that the spots may have no selection-value as far as these species are concerned, and may therefore disappear again in the course of phylogeny, or, on the other hand, that they may be changed in another direction, for instance towards imitation of the rust-red fungoid patches on poplar and willow leaves. In any case we may regard the smallest spots as the initial stages of variation, the larger as a cumulative summation of these. Therefore either these initial stages must already possess selection-value, or, as I said before: there must be some other reason for their cumulative summation. I should like to give one more example, in which we can infer, though we cannot directly observe, the initial stages.

All the Holothurians or sea-cucumbers have in the skin calcareous bodies of different forms, usually thick and irregular, which make the skin tough and resistant. In a small group of them—the species of Synapta—the calcareous bodies occur in the form of delicate anchors of microscopic size. Up till 1897 these anchors, like many other delicate microscopic structures, were regarded as curiosities, as natural marvels. But a Swedish observer, Oestergren, has recently shown that they have a biological significance: they serve the footless Synapta as auxiliary organs of locomotion, since, when the body swells up in the act of creeping, they press firmly with their tips, which are embedded in the skin, against the substratum on which the animal creeps, and thus prevent slipping backwards. In other Holothurians this slipping is made impossible by the fixing of the tube-feet. The anchors act automatically, sinking their tips towards the ground when the corresponding part of the body thickens, and returning to the original position at an angle of 45 degrees to the upper surface when the part becomes thin again. The arms of the anchor do not lie in the same plane as the shaft, and thus the curve of the arms forms the outermost part of the anchor, and offers no further resistance to the gliding of the animal. Every detail of the anchor, the curved portion, the little teeth at the head, the arms, etc., can be interpreted in the most beautiful way, above all the form of the anchor itself, for the two arms prevent it from swaying round to the side. The position of the anchors, too, is definite and significant; they lie obliquely to the longitudinal axis of the animal, and therefore they act alike whether the animal is creeping backwards or forwards. Moreover, the tips would pierce through the skin if the anchors lay in the longitudinal direction. Synapta burrows in the sand; it first pushes in the thin anterior end, and thickens this again, thus enlarging the hole, then the anterior tentacles displace more sand, the body is worked in a little farther, and the process begins anew. In the first act the anchors are passive, but they begin to take an active share in the forward movement when the body is contracted again. Frequently the animal retains only the posterior end buried in the sand, and then the anchors keep it in position, and make rapid withdrawal possible.

Thus we have in these apparently random forms of the calcareous bodies, complex adaptations in which every little detail as to direction, curve, and pointing is exactly determined. That they have selection-value in their present perfected form is beyond all doubt, since the animals are enabled by means of them to bore rapidly into the ground and so to escape from enemies. We do not know what the initial stages were, but we cannot doubt that the little improvements, which occurred as variations of the originally simple slimy bodies of the Holothurians, were preserved because they already possessed selection-value for the Synaptidae. For such minute microscopic structures whose form is so delicately adapted to the role they have to play in the life of the animal, cannot have arisen suddenly and as a whole, and every new variation of the anchor, that is, in the direction of the development of the two arms, and every curving of the shaft which prevented the tips from projecting at the wrong time, in short, every little adaptation in the modelling of the anchor must have possessed selection-value. And that such minute changes of form fall within the sphere of fluctuating variations, that is to say, that they occur is beyond all doubt.

In many of the Synaptidae the anchors are replaced by calcareous rods bent in the form of an S, which are said to act in the same way. Others, such as those of the genus Ankyroderma, have anchors which project considerably beyond the skin, and, according to Oestergren, serve "to catch plant-particles and other substances" and so mask the animal. Thus we see that in the Synaptidae the thick and irregular calcareous bodies of the Holothurians have been modified and transformed in various ways in adaptation to the footlessness of these animals, and to the peculiar conditions of their life, and we must conclude that the earlier stages of these changes presented themselves to the processes of selection in the form of microscopic variations. For it is as impossible to think of any origin other than through selection in this case as in the case of the toughness, and the "drip-tips" of tropical leaves. And as these last could not have been produced directly by the beating of the heavy rain-drops upon them, so the calcareous anchors of Synapta cannot have been produced directly by the friction of the sand and mud at the bottom of the sea, and, since they are parts whose function is PASSIVE the Lamarckian factor of use and disuse does not come into question. The conclusion is unavoidable, that the microscopically small variations of the calcareous bodies in the ancestral forms have been intensified and accumulated in a particular direction, till they have led to the formation of the anchor. Whether this has taken place by the action of natural selection alone, or whether the laws of variation and the intimate processes within the germ-plasm have cooperated will become clear in the discussion of germinal selection. This whole process of adaptation has obviously taken place within the time that has elapsed since this group of sea-cucumbers lost their tube-feet, those characteristic organs of locomotion which occur in no group except the Echinoderms, and yet have totally disappeared in the Synaptidae. And after all what would animals that live in sand and mud do with tube-feet?

(c) Coadaptation.

Darwin pointed out that one of the essential differences between artificial and natural selection lies in the fact that the former can modify only a few characters, usually only one at a time, while Nature preserves in the struggle for existence all the variations of a species, at the same time and in a purely mechanical way, if they possess selection-value.

Herbert Spencer, though himself an adherent of the theory of selection, declared in the beginning of the nineties that in his opinion the range of this principle was greatly over-estimated, if the great changes which have taken place in so many organisms in the course of ages are to be interpreted as due to this process of selection alone, since no transformation of any importance can be evolved by itself; it is always accompanied by a host of secondary changes. He gives the familiar example of the Giant Stag of the Irish peat, the enormous antlers of which required not only a much stronger skull cap, but also greater strength of the sinews, muscles, nerves and bones of the whole anterior half of the animal, if their mass was not to weigh down the animal altogether. It is inconceivable, he says, that so many processes of selection should take place simultaneously, and we are therefore forced to fall back on the Lamarckian factor of the use and disuse of functional parts. And how, he asks, could natural selection follow two opposite directions of evolution in different parts of the body at the same time, as for instance in the case of the kangaroo, in which the forelegs must have become shorter, while the hind legs and the tail were becoming longer and stronger?

Spencer's main object was to substantiate the validity of the Lamarckian principle, the cooperation of which with selection had been doubted by many. And it does seem as though this principle, if it operates in nature at all, offers a ready and simple explanation of all such secondary variations. Not only muscles, but nerves, bones, sinews, in short all tissues which function actively, increase in strength in proportion as they are used, and conversely they decrease when the claims on them diminish. All the parts, therefore, which depend on the part that varied first, as for instance the enlarged antlers of the Irish Elk, must have been increased or decreased in strength, in exact proportion to the claims made upon them,—just as is actually the case.

But beautiful as this explanation would be, I regard it as untenable, because it assumes the transmissibility of functional modifications (so-called "acquired" characters), and this is not only undemonstrable, but is scarcely theoretically conceivable, for the secondary variations which accompany or follow the first as correlative variations, occur also in cases in which the animals concerned are sterile and therefore cannot transmit anything to their descendants. This is true of worker bees, and particularly of ants, and I shall here give a brief survey of the present state of the problem as it appears to me.

Much has been written on both sides of this question since the published controversy on the subject in the nineties between Herbert Spencer and myself. I should like to return to the matter in detail, if the space at my disposal permitted, because it seems to me that the arguments I advanced at that time are equally cogent to-day, notwithstanding all the objections that have since been urged against them. Moreover, the matter is by no means one of subordinate interest; it is the very kernel of the whole question of the reality and value of the principle of selection. For if selection alone does not suffice to explain "harmonious adaptation" as i have called spencer's coadaptation, and if we require to call in the aid of the Lamarckian factor it would be questionable whether selection could explain any adaptations whatever. In this particular case—of worker bees—the Lamarckian factor may be excluded altogether, for it can be demonstrated that here at any rate the effects of use and disuse cannot be transmitted.

But if it be asked why we are unwilling to admit the cooperation of the Darwinian factor of selection and the Lamarckian factor, since this would afford us an easy and satisfactory explanation of the phenomena, I answer:

because the lamarckian principle is fallacious, and because by accepting it we close the way towards deeper insight. It is not a spirit of combativeness or a desire for self-vindication that induces me to take the field once more against the Lamarckian principle, it is the conviction that the progress of our knowledge is being obstructed by the acceptance of this fallacious principle, since the facile explanation it apparently affords prevents our seeking after a truer explanation and a deeper analysis.

The workers in the various species of ants are sterile, that is to say, they take no regular part in the reproduction of the species, although individuals among them may occasionally lay eggs. In addition to this they have lost the wings, and the receptaculum seminis, and their compound eyes have degenerated to a few facets. How could this last change have come about through disuse, since the eyes of workers are exposed to light in the same way as are those of the sexual insects and thus in this particular case are not liable to "disuse" at all? The same is true of the receptaculum seminis, which can only have been disused as far as its glandular portion and its stalk are concerned, and also of the wings, the nerves tracheae and epidermal cells of which could not cease to function until the whole wing had degenerated, for the chitinous skeleton of the wing does not function at all in the active sense.

But, on the other hand, the workers in all species have undergone modifications in a positive direction, as, for instance, the greater development of brain. In many species large workers have evolved,—the so-called soldiers, with enormous jaws and teeth, which defend the colony,—and in others there are small workers which have taken over other special functions, such as the rearing of the young Aphides. This kind of division of the workers into two castes occurs among several tropical species of ants, but it is also present in the Italian species, Colobopsis truncata. Beautifully as the size of the jaws could be explained as due to the increased use made of them by the "soldiers," or the enlarged brain as due to the mental activities of the workers, the fact of the infertility of these forms is an insurmountable obstacle to accepting such an explanation. Neither jaws nor brain can have been evolved on the Lamarckian principle.

The problem of coadaptation is no easier in the case of the ant than in the case of the Giant Stag. Darwin himself gave a pretty illustration to show how imposing the difference between the two kinds of workers in one species would seem if we translated it into human terms. In regard to the Driver ants (Anomma) we must picture to ourselves a piece of work, "for instance the building of a house, being carried on by two kinds of workers, of which one group was five feet four inches high, the other sixteen feet high." ("Origin of Species" (6th edition), page 232.)

Although the ant is a small animal as compared with man or with the Irish Elk, the "soldier" with its relatively enormous jaws is hardly less heavily burdened than the Elk with its antlers, and in the ant's case, too, a strengthening of the skeleton, of the muscles, the nerves of the head, and of the legs must have taken place parallel with the enlargement of the jaws. harmonious adaptation (coadaptation) has here been active in a high degree, and yet these "soldiers" are sterile! There thus remains nothing for it but to refer all their adaptations, positive and negative alike, to processes of selection which have taken place in the rudiments of the workers within the egg and sperm-cells of their parents. There is no way out of the difficulty except the one Darwin pointed out. He himself did not find the solution of the riddle at once. At first he believed that the case of the workers among social insects presented "the most serious special difficulty" in the way of his theory of natural selection; and it was only after it had become clear to him, that it was not the sterile insects themselves but their parents that were selected, according as they produced more or less well adapted workers, that he was able to refer to this very case of the conditions among ants "in order to show the power of natural selection" ("Origin of Species", page 233; see also edition 1, page 242.). He explains his view by a simple but interesting illustration. Gardeners have produced, by means of long continued artificial selection, a variety of Stock, which bears entirely double, and therefore infertile flowers (Ibid. page 230.). Nevertheless the variety continues to be reproduced from seed, because in addition to the double and infertile flowers, the seeds always produce a certain number of single, fertile blossoms, and these are used to reproduce the double variety. These single and fertile plants correspond "to the males and females of an ant-colony, the infertile plants, which are regularly produced in large numbers, to the neuter workers of the colony."

This illustration is entirely apt, the only difference between the two cases consisting in the fact that the variation in the flower is not a useful, but a disadvantageous one, which can only be preserved by artificial selection on the part of the gardener, while the transformations that have taken place parallel with the sterility of the ants are useful, since they procure for the colony an advantage in the struggle for existence, and they are therefore preserved by natural selection. Even the sterility itself in this case is not disadvantageous, since the fertility of the true females has at the same time considerably increased. We may therefore regard the sterile forms of ants, which have gradually been adapted in several directions to varying functions, as a certain proof that selection really takes place in the germ-cells of the fathers and mothers of the workers, and that special complexes of primordia (IDS) are present in the workers and in the males and females, and these complexes contain the primordia of the individual parts (determinants). But since all living entities vary, the determinants must also vary, now in a favourable, now in an unfavourable direction. If a female produces eggs, which contain favourably varying determinants in the worker-ids, then these eggs will give rise to workers modified in the favourable direction, and if this happens with many females, the colony concerned will contain a better kind of worker than other colonies.

I digress here in order to give an account of the intimate processes, which, according to my view, take place within the germ-plasm, and which I have called "germinal selection." These processes are of importance since they form the roots of variation, which in its turn is the root of natural selection. I cannot here do more than give a brief outline of the theory in order to show how the Darwin-Wallace theory of selection has gained support from it.

With others, I regard the minimal amount of substance which is contained within the nucleus of the germ-cells, in the form of rods, bands, or granules, as the germ-substance or germ-plasm, and I call the individual granules IDS. There is always a multiplicity of such ids present in the nucleus, either occurring individually, or united in the form of rods or bands (chromosomes). Each id contains the primary constituents of a whole individual, so that several ids are concerned in the development of a new individual.

In every being of complex structure thousands of primary constituents must go to make up a single id; these I call determinants, and I mean by this name very small individual particles, far below the limits of microscopic visibility, vital units which feed, grow, and multiply by division. These determinants control the parts of the developing embryo,—in what manner need not here concern us. The determinants differ among themselves, those of a muscle are differently constituted from those of a nerve-cell or a glandular cell, etc., and every determinant is in its turn made up of minute vital units, which I call biophors, or the bearers of life. According to my view, these determinants not only assimilate, like every other living unit, but they vary in the course of their growth, as every living unit does; they may vary qualitatively if the elements of which they are composed vary, they may grow and divide more or less rapidly, and their variations give rise to corresponding variations of the organ, cell, or cell-group which they determine. That they are undergoing ceaseless fluctuations in regard to size and quality seems to me the inevitable consequence of their unequal nutrition; for although the germ-cell as a whole usually receives sufficient nutriment, minute fluctuations in the amount carried to different parts within the germ-plasm cannot fail to occur.

Now, if a determinant, for instance of a sensory cell, receives for a considerable time more abundant nutriment than before, it will grow more rapidly—become bigger, and divide more quickly, and, later, when the id concerned develops into an embryo, this sensory cell will become stronger than in the parents, possibly even twice as strong. This is an instance of a hereditary individual variation, arising from the germ.

The nutritive stream which, according to our hypothesis, favours the determinant N by chance, that is, for reasons unknown to us, may remain strong for a considerable time, or may decrease again; but even in the latter case it is conceivable that the ascending movement of the determinant may continue, because the strengthened determinant now actively nourishes itself more abundantly,—that is to say, it attracts the nutriment to itself, and to a certain extent withdraws it from its fellow-determinants. In this way, it may—as it seems to me—get into permanent upward movement, and attain a degree of strength from which there is no falling back. Then positive or negative selection sets in, favouring the variations which are advantageous, setting aside those which are disadvantageous.

In a similar manner a downward variation of the determinants may take place, if its progress be started by a diminished flow of nutriment. The determinants which are weakened by this diminished flow will have less affinity for attracting nutriment because of their diminished strength, and they will assimilate more feebly and grow more slowly, unless chance streams of nutriment help them to recover themselves. But, as will presently be shown, a change of direction cannot take place at every stage of the degenerative process. If a certain critical stage of downward progress be passed, even favourable conditions of food-supply will no longer suffice permanently to change the direction of the variation. Only two cases are conceivable; if the determinant corresponds to a useful organ, only its removal can bring back the germ-plasm to its former level; therefore personal selection removes the id in question, with its determinants, from the germ-plasm, by causing the elimination of the individual in the struggle for existence. But there is another conceivable case; the determinants concerned may be those of an organ which has become useless, and they will then continue unobstructed, but with exceeding slowness, along the downward path, until the organ becomes vestigial, and finally disappears altogether.

The fluctuations of the determinants hither and thither may thus be transformed into a lasting ascending or descending movement; and this is the crucial point of these germinal processes.

This is not a fantastic assumption; we can read it in the fact of the degeneration of disused parts. Useless organs are the only ones which are not helped to ascend again by personal selection, and therefore in their case alone can we form any idea of how the primary constituents behave, when they are subject solely to intra-germinal forces.

The whole determinant system of an id, as I conceive it, is in a state of continual fluctuation upwards and downwards. In most cases the fluctuations will counteract one another, because the passive streams of nutriment soon change, but in many cases the limit from which a return is possible will be passed, and then the determinants concerned will continue to vary in the same direction, till they attain positive or negative selection-value. At this stage personal selection intervenes and sets aside the variation if it is disadvantageous, or favours— that is to say, preserves—it if it is advantageous. Only the determinant of a useless organ is uninfluenced by personal selection, and, as experience shows, it sinks downwards; that is, the organ that corresponds to it degenerates very slowly but uninterruptedly till, after what must obviously be an immense stretch of time, it disappears from the germ-plasm altogether.

Thus we find in the fact of the degeneration of disused parts the proof that not all the fluctuations of a determinant return to equilibrium again, but that, when the movement has attained to a certain strength, it continues in the same direction. We have entire certainty in regard to this as far as the downward progress is concerned, and we must assume it also in regard to ascending variations, as the phenomena of artificial selection certainly justify us in doing. If the Japanese breeders were able to lengthen the tail feathers of the cock to six feet, it can only have been because the determinants of the tail-feathers in the germ-plasm had already struck out a path of ascending variation, and this movement was taken advantage of by the breeder, who continually selected for reproduction the individuals in which the ascending variation was most marked. For all breeding depends upon the unconscious selection of germinal variations.

Of course these germinal processes cannot be proved mathematically, since we cannot actually see the play of forces of the passive fluctuations and their causes. We cannot say how great these fluctuations are, and how quickly or slowly, how regularly or irregularly they change. Nor do we know how far a determinant must be strengthened by the passive flow of the nutritive stream if it is to be beyond the danger of unfavourable variations, or how far it must be weakened passively before it loses the power of recovering itself by its own strength. It is no more possible to bring forward actual proofs in this case than it was in regard to the selection-value of the initial stages of an adaptation. But if we consider that all heritable variations must have their roots in the germ-plasm, and further, that when personal selection does not intervene, that is to say, in the case of parts which have become useless, a degeneration of the part, and therefore also of its determinant must inevitably take place; then we must conclude that processes such as I have assumed are running their course within the germ-plasm, and we can do this with as much certainty as we were able to infer, from the phenomena of adaptation, the selection-value of their initial stages. The fact of the degeneration of disused parts seems to me to afford irrefutable proof that the fluctuations within the germ-plasm are the real root of all hereditary variation, and the preliminary condition for the occurrence of the Darwin-Wallace factor of selection. Germinal selection supplies the stones out of which personal selection builds her temples and palaces: adaptations. The importance for the theory of the process of degeneration of disused parts cannot be over-estimated, especially when it occurs in sterile animal forms, where we are free from the doubt as to the alleged Lamarckian factor which is apt to confuse our ideas in regard to other cases.

If we regard the variation of the many determinants concerned in the transformation of the female into the sterile worker as having come about through the gradual transformation of the ids into worker-ids, we shall see that the germ-plasm of the sexual ants must contain three kinds of ids, male, female, and worker ids, or if the workers have diverged into soldiers and nest-builders, then four kinds. We understand that the worker-ids arose because their determinants struck out a useful path of variation, whether upward or downward, and that they continued in this path until the highest attainable degree of utility of the parts determined was reached. But in addition to the organs of positive or negative selection-value, there were some which were indifferent as far as the success and especially the functional capacity of the workers was concerned: wings, ovarian tubes, receptaculum seminis, a number of the facets of the eye, perhaps even the whole eye. As to the ovarian tubes it is possible that their degeneration was an advantage for the workers, in saving energy, and if so selection would favour the degeneration; but how could the presence of eyes diminish the usefulness of the workers to the colony? or the minute receptaculum seminis, or even the wings? These parts have therefore degenerated because they were of no further value to the insect. But if selection did not influence the setting aside of these parts because they were neither of advantage nor of disadvantage to the species, then the Darwinian factor of selection is here confronted with a puzzle which it cannot solve alone, but which at once becomes clear when germinal selection is added. For the determinants of organs that have no further value for the organism, must, as we have already explained, embark on a gradual course of retrograde development.

In ants the degeneration has gone so far that there are no wing-rudiments present in any species, as is the case with so many butterflies, flies, and locusts, but in the larvae the imaginal discs of the wings are still laid down. With regard to the ovaries, degeneration has reached different levels in different species of ants, as has been shown by the researches of my former pupil, Elizabeth Bickford. In many species there are twelve ovarian tubes, and they decrease from that number to one; indeed, in one species no ovarian tube at all is present. So much at least is certain from what has been said, that in this case everything depends on the fluctuations of the elements of the germ-plasm. Germinal selection, here as elsewhere, presents the variations of the determinants, and personal selection favours or rejects these, or,—if it be a question of organs which have become useless,—it does not come into play at all, and allows the descending variation free course.

It is obvious that even the problem of coadaptation in sterile animals can thus be satisfactorily explained. If the determinants are oscillating upwards and downwards in continual fluctuation, and varying more pronouncedly now in one direction now in the other, useful variations of every determinant will continually present themselves anew, and may, in the course of generations, be combined with one another in various ways. But there is one character of the determinants that greatly facilitates this complex process of selection, that, after a certain limit has been reached, they go on varying in the same direction. From this it follows that development along a path once struck out may proceed without the continual intervention of personal selection. This factor only operates, so to speak, at the beginning, when it selects the determinants which are varying in the right direction, and again at the end, when it is necessary to put a check upon further variation. In addition to this, enormously long periods have been available for all these adaptations, as the very gradual transition stages between females and workers in many species plainly show, and thus this process of transformation loses the marvellous and mysterious character that seemed at the first glance to invest it, and takes rank, without any straining, among the other processes of selection. It seems to me that, from the facts that sterile animal forms can adapt themselves to new vital functions, their superfluous parts degenerate, and the parts more used adapt themselves in an ascending direction, those less used in a descending direction, we must draw the conclusion that harmonious adaptation here comes about without the cooperation of the Lamarckian principle. This conclusion once established, however, we have no reason to refer the thousands of cases of harmonious adaptation, which occur in exactly the same way among other animals or plants, to a principle, the active intervention of which in the transformation of species is nowhere proved. we do not require it to explain the facts, and therefore we must not assume it.

The fact of coadaptation, which was supposed to furnish the strongest argument against the principle of selection, in reality yields the clearest evidence in favour of it. We must assume it, because no other possibility of explanation is open to us, and because these adaptations actually exist, that is to say, have really taken place. With this conviction I attempted, as far back as 1894, when the idea of germinal selection had not yet occurred to me, to make "harmonious adaptation" (coadaptation) more easily intelligible in some way or other, and so I was led to the idea, which was subsequently expounded in detail by Baldwin, and Lloyd Morgan, and also by Osborn, and Gulick as organic selection. It seemed to me that it was not necessary that all the germinal variations required for secondary variations should have occurred simultaneously, since, for instance, in the case of the stag, the bones, muscles, sinews, and nerves would be incited by the increasing heaviness of the antlers to greater activity in the individual life, and so would be strengthened. The antlers can only have increased in size by very slow degrees, so that the muscles and bones may have been able to keep pace with their growth in the individual life, until the requisite germinal variations presented themselves. In this way a disharmony between the increasing weight of the antlers and the parts which support and move them would be avoided, since time would be given for the appropriate germinal variations to occur, and so to set agoing the hereditary variation of the muscles, sinews, and bones. ("The Effect of External Influences upon Development", Romanes Lecture, Oxford, 1894.)

I still regard this idea as correct, but I attribute less importance to "organic selection" than I did at that time, in so far that I do not believe that it ALONE could effect complex harmonious adaptations. Germinal selection now seems to me to play the chief part in bringing about such adaptations. Something the same is true of the principle I have called "Panmixia". As I became more and more convinced, in the course of years, that the Lamarckian principle ought not to be called in to explain the dwindling of disused parts, I believed that this process might be simply explained as due to the cessation of the conservative effect of natural selection. I said to myself that, from the moment in which a part ceases to be of use, natural selection withdraws its hand from it, and then it must inevitably fall from the height of its adaptiveness, because inferior variants would have as good a chance of persisting as better ones, since all grades of fitness of the part in question would be mingled with one another indiscriminately. This is undoubtedly true, as Romanes pointed out ten years before I did, and this mingling of the bad with the good probably does bring about a deterioration of the part concerned. But it cannot account for the steady diminution, which always occurs when a part is in process of becoming rudimentary, and which goes on until it ultimately disappears altogether. The process of dwindling cannot therefore be explained as due to panmixia alone; we can only find a sufficient explanation in germinal selection.

IV. DERIVATIVES OF THE THEORY OF SELECTION

The impetus in all directions given by Darwin through his theory of selection has been an immeasurable one, and its influence is still felt. It falls within the province of the historian of science to enumerate all the ideas which, in the last quarter of the nineteenth century, grew out of Darwin's theories, in the endeavour to penetrate more deeply into the problem of the evolution of the organic world. Within the narrow limits to which this paper is restricted, I cannot attempt to discuss any of these.

V. ARGUMENTS FOR THE REALITY OF THE PROCESSES OF SELECTION

(a) Sexual selection.

Sexual selection goes hand in hand with natural selection. From the very first I have regarded sexual selection as affording an extremely important and interesting corroboration of natural selection, but, singularly enough, it is precisely against this theory that an adverse judgment has been pronounced in so many quarters, and it is only quite recently, and probably in proportion as the wealth of facts in proof of it penetrates into a wider circle, that we seem to be approaching a more general recognition of this side of the problem of adaptation. Thus Darwin's words in his preface to the second edition (1874) of his book, "The Descent of Man and Sexual Selection", are being justified: "My conviction as to the operation of natural selection remains unshaken," and further, "If naturalists were to become more familiar with the idea of sexual selection, it would, I think, be accepted to a much greater extent, and already it is fully and favourably accepted by many competent judges." Darwin was able to speak thus because he was already acquainted with an immense mass of facts, which, taken together, yield overwhelming evidence of the validity of the principle of sexual selection.

Natural selection chooses out for reproduction the individuals that are best equipped for the struggle for existence, and it does so at every stage of development; it thus improves the species in all its stages and forms.

Sexual selection operates only on individuals that are already capable of reproduction, and does so only in relation to the attainment of reproduction. It arises from the rivalry of one sex, usually the male, for the possession of the other, usually the female. Its influence can therefore only directly affect one sex, in that it equips it better for attaining possession of the other. But the effect may extend indirectly to the female sex, and thus the whole species may be modified, without, however, becoming any more capable of resistance in the struggle for existence, for sexual selection only gives rise to adaptations which are likely to give their possessor the victory over rivals in the struggle for possession of the female, and which are therefore peculiar to the wooing sex: the manifold "secondary sexual characters." The diversity of these characters is so great that I cannot here attempt to give anything approaching a complete treatment of them, but I should like to give a sufficient number of examples to make the principle itself, in its various modes of expression, quite clear.

One of the chief preliminary postulates of sexual selection is the unequal number of individuals in the two sexes, for if every male immediately finds his mate there can be no competition for the possession of the female. Darwin has shown that, for the most part, the inequality between the sexes is due simply to the fact that there are more males than females, and therefore the males must take some pains to secure a mate. But the inequality does not always depend on the numerical preponderance of the males, it is often due to polygamy; for, if one male claims several females, the number of females in proportion to the rest of the males will be reduced. Since it is almost always the males that are the wooers, we must expect to find the occurrence of secondary sexual characters chiefly among them, and to find it especially frequent in polygamous species. And this is actually the case.

If we were to try to guess—without knowing the facts—what means the male animals make use of to overcome their rivals in the struggle for the possession of the female, we might name many kinds of means, but it would be difficult to suggest any which is not actually employed in some animal group or other. I begin with the mere difference in strength, through which the male of many animals is so sharply distinguished from the female, as, for instance, the lion, walrus, "sea-elephant," and others. Among these the males fight violently for the possession of the female, who falls to the victor in the combat. In this simple case no one can doubt the operation of selection, and there is just as little room for doubt as to the selection-value of the initial stages of the variation. Differences in bodily strength are apparent even among human beings, although in their case the struggle for the possession of the female is no longer decided by bodily strength alone.

Combats between male animals are often violent and obstinate, and the employment of the natural weapons of the species in this way has led to perfecting of these, e.g. the tusks of the boar, the antlers of the stag, and the enormous, antler-like jaws of the stag-beetle. Here again it is impossible to doubt that variations in these organs presented themselves, and that these were considerable enough to be decisive in combat, and so to lead to the improvement of the weapon.

Among many animals, however, the females at first withdraw from the males; they are coy, and have to be sought out, and sometimes held by force. This tracking and grasping of the females by the males has given rise to many different characters in the latter, as, for instance, the larger eyes of the male bee, and especially of the males of the Ephemerids (May-flies), some species of which show, in addition to the usual compound eyes, large, so-called turban-eyes, so that the whole head is covered with seeing surfaces. In these species the females are very greatly in the minority (1-100), and it is easy to understand that a keen competition for them must take place, and that, when the insects of both sexes are floating freely in the air, an unusually wide range of vision will carry with it a decided advantage. Here again the actual adaptations are in accordance with the preliminary postulates of the theory. We do not know the stages through which the eye has passed to its present perfected state, but, since the number of simple eyes (facets) has become very much greater in the male than in the female, we may assume that their increase is due to a gradual duplication of the determinants of the ommatidium in the germ-plasm, as I have already indicated in regard to sense-organs in general. In this case, again, the selection-value of the initial stages hardly admits of doubt; better vision directly secures reproduction.

In many cases the organ of smell shows a similar improvement. Many lower Crustaceans (Daphnidae) have better developed organs of smell in the male sex. The difference is often slight and amounts only to one or two olfactory filaments, but certain species show a difference of nearly a hundred of these filaments (Leptodora). The same thing occurs among insects.

We must briefly consider the clasping or grasping organs which have developed in the males among many lower Crustaceans, but here natural selection plays its part along with sexual selection, for the union of the sexes is an indispensable condition for the maintenance of the species, and as Darwin himself pointed out, in many cases the two forms of selection merge into each other. This fact has always seemed to me to be a proof of natural selection, for, in regard to sexual selection, it is quite obvious that the victory of the best-equipped could have brought about the improvement only of the organs concerned, the factors in the struggle, such as the eye and the olfactory organ.

We come now to the excitants; that is, to the group of sexual characters whose origin through processes of selection has been most frequently called in question. We may cite the love-calls produced by many male insects, such as crickets and cicadas. These could only have arisen in animal groups in which the female did not rapidly flee from the male, but was inclined to accept his wooing from the first. Thus, notes like the chirping of the male cricket serve to entice the females. At first they were merely the signal which showed the presence of a male in the neighbourhood, and the female was gradually enticed nearer and nearer by the continued chirping. The male that could make himself heard to the greatest distance would obtain the largest following, and would transmit the beginnings, and, later, the improvement of his voice to the greatest number of descendants. But sexual excitement in the female became associated with the hearing of the love-call, and then the sound-producing organ of the male began to improve, until it attained to the emission of the long-drawn-out soft notes of the mole-cricket or the maenad-like cry of the cicadas. I cannot here follow the process of development in detail, but will call attention to the fact that the original purpose of the voice, the announcing of the male's presence, became subsidiary, and the exciting of the female became the chief goal to be aimed at. The loudest singers awakened the strongest excitement, and the improvement resulted as a matter of course. I conceive of the origin of bird-song in a somewhat similar manner, first as a means of enticing, then of exciting the female.

One more kind of secondary sexual character must here be mentioned: the odour which emanates from so many animals at the breeding season. It is possible that this odour also served at first merely to give notice of the presence of individuals of the other sex, but it soon became an excitant, and as the individuals which caused the greatest degree of excitement were preferred, it reached as high a pitch of perfection as was possible to it. I shall confine myself here to the comparatively recently discovered fragrance of butterflies. Since Fritz Muller found out that certain Brazilian butterflies gave off fragrance "like a flower," we have become acquainted with many such cases, and we now know that in all lands, not only many diurnal Lepidoptera but nocturnal ones also give off a delicate odour, which is agreeable even to man. The ethereal oil to which this fragrance is due is secreted by the skin-cells, usually of the wing, as I showed soon after the discovery of the scent-scales. This is the case in the males; the females have no special scent-scales recognisable as such by their form, but they must, nevertheless, give off an extremely delicate fragrance, although our imperfect organ of smell cannot perceive it, for the males become aware of the presence of a female, even at night, from a long distance off, and gather round her. We may therefore conclude, that both sexes have long given forth a very delicate perfume, which announced their presence to others of the same species, and that in many species (not in all) these small beginnings became, in the males, particularly strong scent-scales of characteristic form (lute, brush, or lyre-shaped). At first these scales were scattered over the surface of the wing, but gradually they concentrated themselves, and formed broad, velvety bands, or strong, prominent brushes, and they attained their highest pitch of evolution when they became enclosed within pits or folds of the skin, which could be opened to let the delicious fragrance stream forth suddenly towards the female. Thus in this case also we see that characters, the original use of which was to bring the sexes together, and so to maintain the species, have been evolved in the males into means for exciting the female. And we can hardly doubt, that the females are most readily enticed to yield to the butterfly that sends out the strongest fragrance,—that is to say, that excites them to the highest degree. It is a pity that our organs of smell are not fine enough to examine the fragrance of male Lepidoptera in general, and to compare it with other perfumes which attract these insects. (See Poulton, "Essays on Evolution", 1908, pages 316, 317.) As far as we can perceive them they resemble the fragrance of flowers, but there are Lepidoptera whose scent suggests musk. A smell of musk is also given off by several plants: it is a sexual excitant in the musk-deer, the musk-sheep, and the crocodile.

As far as we know, then, it is perfumes similar to those of flowers that the male Lepidoptera give off in order to entice their mates, and this is a further indication that animals, like plants, can to a large extent meet the claims made upon them by life, and produce the adaptations which are most purposive,—a further proof, too, of my proposition that the useful variations, so to speak, are always there. The flowers developed the perfumes which entice their visitors, and the male Lepidoptera developed the perfumes which entice and excite their mates.

There are many pretty little problems to be solved in this connection, for there are insects, such as some flies, that are attracted by smells which are unpleasant to us, like those from decaying flesh and carrion. But there are also certain flowers, some orchids for instance, which give forth no very agreeable odour, but one which is to us repulsive and disgusting; and we should therefore expect that the males of such insects would give off a smell unpleasant to us, but there is no case known to me in which this has been demonstrated.

In cases such as we have discussed, it is obvious that there is no possible explanation except through selection. This brings us to the last kind of secondary sexual characters, and the one in regard to which doubt has been most frequently expressed,—decorative colours and decorative forms, the brilliant plumage of the male pheasant, the humming-birds, and the bird of Paradise, as well as the bright colours of many species of butterfly, from the beautiful blue of our little Lycaenidae to the magnificent azure of the large Morphinae of Brazil. In a great many cases, though not by any means in all, the male butterflies are "more beautiful" than the females, and in the Tropics in particular they shine and glow in the most superb colours. I really see no reason why we should doubt the power of sexual selection, and I myself stand wholly on Darwin's side. Even though we certainly cannot assume that the females exercise a conscious choice of the "handsomest" mate, and deliberate like the judges in a court of justice over the perfections of their wooers, we have no reason to doubt that distinctive forms (decorative feathers) and colours have a particularly exciting effect upon the female, just as certain odours have among animals of so many different groups, including the butterflies. The doubts which existed for a considerable time, as a result of fallacious experiments, as to whether the colours of flowers really had any influence in attracting butterflies have now been set at rest through a series of more careful investigations; we now know that the colours of flowers are there on account of the butterflies, as Sprengel first showed, and that the blossoms of Phanerogams are selected in relation to them, as Darwin pointed out.

Certainly it is not possible to bring forward any convincing proof of the origin of decorative colours through sexual selection, but there are many weighty arguments in favour of it, and these form a body of presumptive evidence so strong that it almost amounts to certainty.

In the first place, there is the analogy with other secondary sexual characters. If the song of birds and the chirping of the cricket have been evolved through sexual selection, if the penetrating odours of male animals,—the crocodile, the musk-deer, the beaver, the carnivores, and, finally, the flower-like fragrances of the butterflies have been evolved to their present pitch in this way, why should decorative colours have arisen in some other way? Why should the eye be less sensitive to specifically male colours and other visible signs enticing to the female, than the olfactory sense to specifically male odours, or the sense of hearing to specifically male sounds? Moreover, the decorative feathers of birds are almost always spread out and displayed before the female during courtship. I have elsewhere ("The Evolution Theory", London, 1904, I. page 219.) pointed out that decorative colouring and sweet-scentedness may replace one another in Lepidoptera as well as in flowers, for just as some modestly coloured flowers (mignonette and violet) have often a strong perfume, while strikingly coloured ones are sometimes quite devoid of fragrance, so we find that the most beautiful and gaily-coloured of our native Lepidoptera, the species of Vanessa, have no scent-scales, while these are often markedly developed in grey nocturnal Lepidoptera. Both attractions may, however, be combined in butterflies, just as in flowers. Of course, we cannot explain why both means of attraction should exist in one genus, and only one of them in another, since we do not know the minutest details of the conditions of life of the genera concerned. But from the sporadic distribution of scent-scales in Lepidoptera, and from their occurrence or absence in nearly related species, we may conclude that fragrance is a relatively modern acquirement, more recent than brilliant colouring.

One thing in particular that stamps decorative colouring as a product of selection is its gradual intensification by the addition of new spots, which we can quite well observe, because in many cases the colours have been first acquired by the males, and later transmitted to the females by inheritance. The scent-scales are never thus transmitted, probably for the same reason that the decorative colours of many birds are often not transmitted to the females: because with these they would be exposed to too great elimination by enemies. Wallace was the first to point out that in species with concealed nests the beautiful feathers of the male occurred in the female also, as in the parrots, for instance, but this is not the case in species which brood on an exposed nest. In the parrots one can often observe that the general brilliant colouring of the male is found in the female, but that certain spots of colour are absent, and these have probably been acquired comparatively recently by the male and have not yet been transmitted to the female.

Isolation of the group of individuals which is in process of varying is undoubtedly of great value in sexual selection, for even a solitary conspicuous variation will become dominant much sooner in a small isolated colony, than among a large number of members of a species.

Anyone who agrees with me in deriving variations from germinal selection will regard that process as an essential aid towards explaining the selection of distinctive courtship-characters, such as coloured spots, decorative feathers, horny outgrowths in birds and reptiles, combs, feather-tufts, and the like, since the beginnings of these would be presented with relative frequency in the struggle between the determinants within the germ-plasm. The process of transmission of decorative feathers to the female results, as Darwin pointed out and illustrated by interesting examples, in the colour-transformation of a whole species, and this process, as the phyletically older colouring of young birds shows, must, in the course of thousands of years, have repeated itself several times in a line of descent.

If we survey the wealth of phenomena presented to us by secondary sexual characters, we can hardly fail to be convinced of the truth of the principle of sexual selection. And certainly no one who has accepted natural selection should reject sexual selection, for, not only do the two processes rest upon the same basis, but they merge into one another, so that it is often impossible to say how much of a particular character depends on one and how much on the other form of selection.

(b) Natural Selection

An actual proof of the theory of sexual selection is out of the question, if only because we cannot tell when a variation attains to selection-value. It is certain that a delicate sense of smell is of value to the male moth in his search for the female, but whether the possession of one additional olfactory hair, or of ten, or of twenty additional hairs leads to the success of its possessor we are unable to tell. And we are groping even more in the dark when we discuss the excitement caused in the female by agreeable perfumes, or by striking and beautiful colours. That these do make an impression is beyond doubt; but we can only assume that slight intensifications of them give any advantage, and we must assume this since otherwise secondary sexual characters remain inexplicable.

The same thing is true in regard to natural selection. It is not possible to bring forward any actual proof of the selection-value of the initial stages, and the stages in the increase of variations, as has been already shown. But the selection-value of a finished adaptation can in many cases be statistically determined. Cesnola and Poulton have made valuable experiments in this direction. The former attached forty-five individuals of the green, and sixty-five of the brown variety of the praying mantis (Mantis religiosa), by a silk thread to plants, and watched them for seventeen days. The insects which were on a surface of a colour similar to their own remained uneaten, while twenty-five green insects on brown parts of plants had all disappeared in eleven days.

The experiments of Poulton and Sanders ("Report of the British Association" (Bristol, 1898), London, 1899, pages 906-909.) were made with 600 pupae of Vanessa urticae, the "tortoise-shell butterfly." The pupae were artificially attached to nettles, tree-trunks, fences, walls, and to the ground, some at Oxford, some at St Helens in the Isle of Wight. In the course of a month 93 per cent of the pupae at Oxford were killed, chiefly by small birds, while at St Helens 68 per cent perished. The experiments showed very clearly that the colour and character of the surface on which the pupa rests—and thus its own conspicuousness—are of the greatest importance. At Oxford only the four pupae which were fastened to nettles emerged; all the rest—on bark, stones and the like—perished. At St Helens the elimination was as follows: on fences where the pupae were conspicuous, 92 per cent; on bark, 66 per cent; on walls, 54 per cent; and among nettles, 57 per cent. These interesting experiments confirm our views as to protective coloration, and show further, that the ratio of elimination in the species is a very high one, and that therefore selection must be very keen.

We may say that the process of selection follows as a logical necessity from the fulfilment of the three preliminary postulates of the theory: variability, heredity, and the struggle for existence, with its enormous ratio of elimination in all species. To this we must add a fourth factor, the intensification of variations which Darwin established as a fact, and which we are now able to account for theoretically on the basis of germinal selection. It may be objected that there is considerable uncertainty about this logical proof, because of our inability to demonstrate the selection-value of the initial stages and the individual stages of increase. We have therefore to fall back on presumptive evidence. This is to be found in the interpretative value of the theory. Let us consider this point in greater detail.

In the first place, it is necessary to emphasise what is often overlooked, namely, that the theory not only explains the transformations of species, it also explains their remaining the same; in addition to the principle of varying, it contains within itself that of persisting. It is part of the essence of selection, that it not only causes a part to vary till it has reached its highest pitch of adaptation, but that it maintains it at this pitch. this conserving influence of natural selection is of great importance, and was early recognised by Darwin; it follows naturally from the principle of the survival of the fittest.

We understand from this how it is that a species which has become fully adapted to certain conditions of life ceases to vary, but remains "constant," as long as the conditions of life for it remain unchanged, whether this be for thousands of years, or for whole geological epochs. But the most convincing proof of the power of the principle of selection lies in the innumerable multitude of phenomena which cannot be explained in any other way. To this category belong all structures which are only passively of advantage to the organism, because none of these can have arisen by the alleged lamarckian principle. These have been so often discussed that we need do no more than indicate them here. Until quite recently the sympathetic coloration of animals—for instance, the whiteness of Arctic animals—was referred, at least in part, to the direct influence of external factors, but the facts can best be explained by referring them to the processes of selection, for then it is unnecessary to make the gratuitous assumption that many species are sensitive to the stimulus of cold and that others are not. The great majority of Arctic land-animals, mammals and birds, are white, and this proves that they were all able to present the variation which was most useful for them. The sable is brown, but it lives in trees, where the brown colouring protects and conceals it more effectively. The musk-sheep (Ovibos moschatus) is also brown, and contrasts sharply with the ice and snow, but it is protected from beasts of prey by its gregarious habit, and therefore it is of advantage to be visible from as great a distance as possible. That so many species have been able to give rise to white varieties does not depend on a special sensitiveness of the skin to the influence of cold, but to the fact that Mammals and Birds have a general tendency to vary towards white. Even with us, many birds—starlings, blackbirds, swallows, etc.—occasionally produce white individuals, but the white variety does not persist, because it readily falls a victim to the carnivores. This is true of white fawns, foxes, deer, etc. The whiteness, therefore, arises from internal causes, and only persists when it is useful. A great many animals living in a green environment have become clothed in green, especially insects, caterpillars, and Mantidae, both persecuted and persecutors.

That it is not the direct effect of the environment which calls forth the green colour is shown by the many kinds of caterpillar which rest on leaves and feed on them, but are nevertheless brown. These feed by night and betake themselves through the day to the trunk of the tree, and hide in the furrows of the bark. We cannot, however, conclude from this that they were unable to vary towards green, for there are Arctic animals which are white only in winter and brown in summer (Alpine hare, and the ptarmigan of the Alps), and there are also green leaf-insects which remain green only while they are young and difficult to see on the leaf, but which become brown again in the last stage of larval life, when they have outgrown the leaf. They then conceal themselves by day, sometimes only among withered leaves on the ground, sometimes in the earth itself. It is interesting that in one genus, Chaerocampa, one species is brown in the last stage of larval life, another becomes brown earlier, and in many species the last stage is not wholly brown, a part remaining green. Whether this is a case of a double adaptation, or whether the green is being gradually crowded out by the brown, the fact remains that the same species, even the same individual, can exhibit both variations. The case is the same with many of the leaf-like Orthoptera, as, for instance, the praying mantis (Mantis religiosa) which we have already mentioned.

But the best proofs are furnished by those often-cited cases in which the insect bears a deceptive resemblance to another object. We now know many such cases, such as the numerous imitations of green or withered leaves, which are brought about in the most diverse ways, sometimes by mere variations in the form of the insect and in its colour, sometimes by an elaborate marking, like that which occurs in the Indian leaf-butterflies, Kallima inachis. In the single butterfly-genus Anaea, in the woods of South America, there are about a hundred species which are all gaily coloured on the upper surface, and on the reverse side exhibit the most delicate imitation of the colouring and pattern of a leaf, generally without any indication of the leaf-ribs, but extremely deceptive nevertheless. Anyone who has seen only one such butterfly may doubt whether many of the insignificant details of the marking can really be of advantage to the insect. Such details are for instance the apparent holes and splits in the apparently dry or half-rotten leaf, which are usually due to the fact that the scales are absent on a circular or oval patch so that the colourless wing-membrane lies bare, and one can look through the spot as through a window. Whether the bird which is seeking or pursuing the butterflies takes these holes for dewdrops, or for the work of a devouring insect, does not affect the question; the mirror-like spot undoubtedly increases the general deceptiveness, for the same thing occurs in many leaf-butterflies, though not in all, and in some cases it is replaced in quite a peculiar manner. In one species of Anaea (A. divina), the resting butterfly looks exactly like a leaf out of the outer edge of which a large semicircular piece has been eaten, possibly by a caterpillar; but if we look more closely it is obvious that there is no part of the wing absent, and that the semicircular piece is of a clear, pale yellow colour, while the rest of the wing is of a strongly contrasted dark brown.

But the deceptive resemblance may be caused in quite a different manner. I have often speculated as to what advantage the brilliant white C could give to the otherwise dusky-coloured "Comma butterfly" (Grapta C. album). Poulton's recent observations ("Proc. Ent. Soc"., London, May 6, 1903.) have shown that this represents the imitation of a crack such as is often seen in dry leaves, and is very conspicuous because the light shines through it.

The utility obviously lies in presenting to the bird the very familiar picture of a broken leaf with a clear shining slit, and we may conclude, from the imitation of such small details, that the birds are very sharp observers and that the smallest deviation from the usual arrests their attention and incites them to closer investigation. It is obvious that such detailed—we might almost say such subtle— deceptive resemblances could only have come about in the course of long ages through the acquirement from time to time of something new which heightened the already existing resemblance.

In face of facts like these there can be no question of chance, and no one has succeeded so far in finding any other explanation to replace that by selection. For the rest, the apparent leaves are by no means perfect copies of a leaf; many of them only represent the torn or broken piece, or the half or two-thirds of a leaf, but then the leaves themselves frequently do not present themselves to the eye as a whole, but partially concealed among other leaves. Even those butterflies which, like the species of Kallima and Anaea, represent the whole of a leaf with stalk, ribs, apex, and the whole breadth, are not actual copies which would satisfy a botanist; there is often much wanting. In Kallima the lateral ribs of the leaf are never all included in the markings; there are only two or three on the left side and at most four or five on the right, and in many individuals these are rather obscure, while in others they are comparatively distinct. This furnishes us with fresh evidence in favour of their origin through processes of selection, for a botanically perfect picture could not arise in this way; there could only be a fixing of such details as heightened the deceptive resemblance.

Our postulate of origin through selection also enables us to understand why the leaf-imitation is on the lower surface of the wing in the diurnal Lepidoptera, and on the upper surface in the nocturnal forms, corresponding to the attitude of the wings in the resting position of the two groups.

The strongest of all proofs of the theory, however, is afforded by cases of true "mimicry," those adaptations discovered by Bates in 1861, consisting in the imitation of one species by another, which becomes more and more like its model. The model is always a species that enjoys some special protection from enemies, whether because it is unpleasant to taste, or because it is in some way dangerous.

It is chiefly among insects and especially among butterflies that we find the greatest number of such cases. Several of these have been minutely studied, and every detail has been investigated, so that it is difficult to understand how there can still be disbelief in regard to them. If the many and exact observations which have been carefully collected and critically discussed, for instance by Poulton ("Essays on Evolution", 1889-1907, Oxford, 1908, passim, e.g. page 269.) were thoroughly studied, the arguments which are still frequently urged against mimicry would be found untenable; we can hardly hope to find more convincing proof of the actuality of the processes of selection than these cases put into our hands. The preliminary postulates of the theory of mimicry have been disputed, for instance, that diurnal butterflies are persecuted and eaten by birds, but observations specially directed towards this point in India, Africa, America and Europe have placed it beyond all doubt. If it were necessary I could myself furnish an account of my own observations on this point.

In the same way it has been established by experiment and observation in the field that in all the great regions of distribution there are butterflies which are rejected by birds and lizards, their chief enemies, on account of their unpleasant smell or taste. These butterflies are usually gaily and conspicuously coloured and thus—as Wallace first interpreted it—are furnished with an easily recognisable sign: a sign of unpalatableness or warning colours. If they were not thus recognisable easily and from a distance, they would frequently be pecked at by birds, and then rejected because of their unpleasant taste; but as it is, the insect-eaters recognise them at once as unpalatable booty and ignore them. Such Immune (The expression does not refer to all the enemies of this butterfly; against ichneumon-flies, for instance, their unpleasant smell usually gives no protection.) species, wherever they occur, are imitated by other palatable species, which thus acquire a certain degree of protection.

It is true that this explanation of the bright, conspicuous colours is only a hypothesis, but its foundations,—unpalatableness, and the liability of other butterflies to be eaten,—are certain, and its consequences—the existence of mimetic palatable forms—confirm it in the most convincing manner. Of the many cases now known I select one, which is especially remarkable, and which has been thoroughly investigated, Papilio dardanus (merope), a large, beautiful, diurnal butterfly which ranges from Abyssinia throughout the whole of Africa to the south coast of Cape Colony.

The males of this form are everywhere almost the same in colour and in form of wings, save for a few variations in the sparse black markings on the pale yellow ground. But the females occur in several quite different forms and colourings, and one of these only, the Abyssinian form, is like the male, while the other three or four are mimetic, that is to say, they copy a butterfly of quite a different family the Danaids, which are among the immune forms. In each region the females have thus copied two or three different immune species. There is much that is interesting to be said in regard to these species, but it would be out of keeping with the general tenor of this paper to give details of this very complicated case of polymorphism in P. dardanus. Anyone who is interested in the matter will find a full and exact statement of the case in as far as we know it, in Poulton's "Essays on Evolution" (pages 373-375). (Professor Poulton has corrected some wrong descriptions which I had unfortunately overlooked in the Plates of my book "Vortrage uber Descendenztheorie", and which refer to Papilio dardanus (merope). These mistakes are of no importance as far as and understanding of the mimicry-theory is concerned, but I hope shortly to be able to correct them in a later edition.) I need only add that three different mimetic female forms have been reared from the eggs of a single female in South Africa. The resemblance of these forms to their immune models goes so far that even the details of the local forms of the models are copied by the mimetic species.

It remains to be said that in Madagascar a butterfly, Papilio meriones, occurs, of which both sexes are very similar in form and markings to the non-mimetic male of P. dardanus, so that it probably represents the ancestor of this latter species.

In face of such facts as these every attempt at another explanation must fail. Similarly all the other details of the case fulfil the preliminary postulates of selection, and leave no room for any other interpretation. That the males do not take on the protective colouring is easily explained, because they are in general more numerous, and the females are more important for the preservation of the species, and must also live longer in order to deposit their eggs. We find the same state of things in many other species, and in one case (Elymnias undularis) in which the male is also mimetically coloured, it copies quite a differently coloured immune species from the model followed by the female. This is quite intelligible when we consider that if there were too many false immune types, the birds would soon discover that there were palatable individuals among those with unpalatable warning colours. Hence the imitation of different immune species by Papilio dardanus!

I regret that lack of space prevents my bringing forward more examples of mimicry and discussing them fully. But from the case of Papilio dardanus alone there is much to be learnt which is of the highest importance for our understanding of transformations. It shows us chiefly what I once called, somewhat strongly perhaps, the omnipotence of natural selection in answer to an opponent who had spoken of its "inadequacy." We here see that one and the same species is capable of producing four or five different patterns of colouring and marking; thus the colouring and marking are not, as has often been supposed, a necessary outcome of the specific nature of the species, but a true adaptation, which cannot arise as a direct effect of climatic conditions, but solely through what I may call the sorting out of the variations produced by the species, according to their utility. That caterpillars may be either green or brown is already something more than could have been expected according to the old conception of species, but that one and the same butterfly should be now pale yellow, with black; now red with black and pure white; now deep black with large, pure white spots; and again black with a large ochreous-yellow spot, and many small white and yellow spots; that in one sub-species it may be tailed like the ancestral form, and in another tailless like its Danaid model,—all this shows a far-reaching capacity for variation and adaptation that wide never have expected if we did not see the facts before us. How it is possible that the primary colour-variations should thus be intensified and combined remains a puzzle even now; we are reminded of the modern three-colour printing,—perhaps similar combinations of the primary colours take place in this case; in any case the direction of these primary variations is determined by the artist whom we know as natural selection, for there is no other conceivable way in which the model could affect the butterfly that is becoming more and more like it. The same climate surrounds all four forms of female; they are subject to the same conditions of nutrition. Moreover, Papilio dardanus is by no means the only species of butterfly which exhibits different kinds of colour-pattern on its wings. Many species of the Asiatic genus Elymnias have on the upper surface a very good imitation of an immune Euploeine (Danainae), often with a steel-blue ground-colour, while the under surface is well concealed when the butterfly is at rest,—thus there are two kinds of protective coloration each with a different meaning! The same thing may be observed in many non-mimetic butterflies, for instance in all our species of Vanessa, in which the under side shows a grey-brown or brownish-black protective coloration, but we do not yet know with certainty what may be the biological significance of the gaily coloured upper surface.

In general it may be said that mimetic butterflies are comparatively rare species, but there are exceptions, for instance Limenitis archippus in North America, of which the immune model (Danaida plexippus) also occurs in enormous numbers.

In another mimicry-category the imitators are often more numerous than the models, namely in the case of the imitation of DANGEROUS INSECTS by harmless species. Bees and wasps are dreaded for their sting, and they are copied by harmless flies of the genera Eristalis and Syrphus, and these mimics often occur in swarms about flowering plants without damage to themselves or to their models; they are feared and are therefore left unmolested.

In regard also to the faithfulness of the copy the facts are quite in harmony with the theory, according to which the resemblance must have arisen and increased by degrees. We can recognise this in many cases, for even now the mimetic species show very varying degrees of resemblance to their immune model. If we compare, for instance, the many different imitators of Danaida chrysippus we find that, with their brownish-yellow ground-colour, and the position and size, and more or less sharp limitation of their clear marginal spots, they have reached very different degrees of nearness to their model. Or compare the female of Elymnias undularis with its model Danaida genutia; there is a general resemblance, but the marking of the Danaida is very roughly imitated in Elymnias.

Another fact that bears out the theory of mimicry is, that even when the resemblance in colour-pattern is very great, the wing-venation, which is so constant, and so important in determining the systematic position of butterflies, is never affected by the variation. The pursuers of the butterfly have no time to trouble about entomological intricacies.

I must not pass over a discovery of Poulton's which is of great theoretical importance—that mimetic butterflies may reach the same effect by very different means. ("Journ. Linn. Soc. London (Zool.)", Vol. XXVI. 1898, pages 598-602.) Thus the glass-like transparency of the wing of a certain Ithomiine (Methona) and its Pierine mimic (Dismorphia orise) depends on a diminution in the size of the scales; in the Danaine genus Ituna it is due to the fewness of the scales, and in a third imitator, a moth (Castnia linus var. heliconoides) the glass-like appearance of the wing is due neither to diminution nor to absence of scales, but to their absolute colourlessness and transparency, and to the fact that they stand upright. In another moth mimic (Anthomyza) the arrangement of the transparent scales is normal. Thus it is not some unknown external influence that has brought about the transparency of the wing in these five forms, as has sometimes been supposed. Nor is it a hypothetical internal evolutionary tendency, for all three vary in a different manner. The cause of this agreement can only lie in selection, which preserves and intensifies in each species the favourable variations that present themselves. The great faithfulness of the copy is astonishing in these cases, for it is not the whole wing which is transparent; certain markings are black in colour, and these contrast sharply with the glass-like ground. It is obvious that the pursuers of these butterflies must be very sharp-sighted, for otherwise the agreement between the species could never have been pushed so far. The less the enemies see and observe, the more defective must the imitation be, and if they had been blind, no visible resemblance between the species which required protection could ever have arisen.

A seemingly irreconcilable contradiction to the mimicry theory is presented in the following cases, which were known to Bates, who, however, never succeeded in bringing them into line with the principle of mimicry.

In South America there are, as we have already said, many mimics of the immune Ithomiinae (or as Bates called them Heliconidae). Among these there occur not merely species which are edible, and thus require the protection of a disguise, but others which are rejected on account of their unpalatableness. How could the Ithomiine dress have developed in their case, and of what use is it, since the species would in any case be immune? In Eastern Brazil, for instance, there are four butterflies, which bear a most confusing resemblance to one another in colour, marking, and form of wing, and all four are unpalatable to birds. They belong to four different genera and three sub-families, and we have to inquire: Whence came this resemblance and what end does it serve? For a long time no satisfactory answer could be found, but Fritz Muller (In "Kosmos", 1879, page 100.), seventeen years after Bates, offered a solution to the riddle, when he pointed out that young birds could not have an instinctive knowledge of the unpalatableness of the Ithomiines, but must learn by experience which species were edible and which inedible. Thus each young bird must have tasted at least one individual of each inedible species and discovered its unpalatability, before it learnt to avoid, and thus to spare the species. But if the four species resemble each other very closely the bird will regard them all as of the same kind, and avoid them all. Thus there developed a process of selection which resulted in the survival of the

Ithomiine-like individuals, and in so great an increase of resemblance between the four species, that they are difficult to distinguish one from another even in a collection. The advantage for the four species, living side by side as they do e.g. in Bahia, lies in the fact that only one individual from the mimicry-ring ("inedible association") need be tasted by a young bird, instead of at least four individuals, as would otherwise be the case. As the number of young birds is great, this makes a considerable difference in the ratio of elimination.

These interesting mimicry-rings (trusts), which have much significance for the theory, have been the subject of numerous and careful investigations, and at least their essential features are now fully established. Muller took for granted, without making any investigations, that young birds only learn by experience to distinguish between different kinds of victims. But Lloyd Morgan's ("Habit and Instinct", London, 1896.) experiments with young birds proved that this is really the case, and at the same time furnished an additional argument against the Lamarckian principle.

In addition to the mimicry-rings first observed in South America, others have been described from Tropical India by Moore, and by Poulton and Dixey from Africa, and we may expect to learn many more interesting facts in this connection. Here again the preliminary postulates of the theory are satisfied. And how much more that would lead to the same conclusion might be added!

As in the case of mimicry many species have come to resemble one another through processes of selection, so we know whole classes of phenomena in which plants and animals have become adapted to one another, and have thus been modified to a considerable degree. I refer particularly to the relation between flowers and insects; but as there is an article on "The Biology of Flowers" in this volume, I need not discuss the subject, but will confine myself to pointing out the significance of these remarkable cases for the theory of selection. Darwin has shown that the originally inconspicuous blossoms of the phanerogams were transformed into flowers through the visits of insects, and that, conversely, several large orders of insects have been gradually modified by their association with flowers, especially as regards the parts of their body actively concerned. Bees and butterflies in particular have become what they are through their relation to flowers. In this case again all that is apparently contradictory to the theory can, on closer investigation, be beautifully interpreted in corroboration of it. Selection can give rise only to what is of use to the organism actually concerned, never to what is of use to some other organism, and we must therefore expect to find that in flowers only characters of use to themselves have arisen, never characters which are of use to insects only, and conversely that in the insects characters useful to them and not merely to the plants would have originated. For a long time it seemed as if an exception to this rule existed in the case of the fertilisation of the yucca blossoms by a little moth, Pronuba yuccasella. This little moth has a sickle-shaped appendage to its mouth-parts which occurs in no other Lepidopteron, and which is used for pushing the yellow pollen into the opening of the pistil, thus fertilising the flower. Thus it appears as if a new structure, which is useful only to the plant, has arisen in the insect. But the difficulty is solved as soon as we learn that the moth lays its eggs in the fruit-buds of the Yucca, and that the larvae, when they emerge, feed on the developing seeds. In effecting the fertilisation of the flower the moth is at the same time making provision for its own offspring, since it is only after fertilisation that the seeds begin to develop. There is thus nothing to prevent our referring this structural adaptation in Pronuba yuccasella to processes of selection, which have gradually transformed the maxillary palps of the female into the sickle-shaped instrument for collecting the pollen, and which have at the same time developed in the insect the instinct to press the pollen into the pistil.

In this domain, then, the theory of selection finds nothing but corroboration, and it would be impossible to substitute for it any other explanation, which, now that the facts are so well known, could be regarded as a serious rival to it. That selection is a factor, and a very powerful factor in the evolution of organisms, can no longer be doubted. Even although we cannot bring forward formal proofs of it in detail, cannot calculate definitely the size of the variations which present themselves, and their selection-value, cannot, in short, reduce the whole process to a mathematical formula, yet we must assume selection, because it is the only possible explanation applicable to whole classes of phenomena, and because, on the other hand, it is made up of factors which we know can be proved actually to exist, and which, IF they exist, must of logical necessity cooperate in the manner required by the theory. We must accept it because the phenomena of evolution and adaptation must have a natural basis, and because it is the only possible explanation of them. (This has been discussed in many of my earlier works. See for instance "The All-Sufficiency of Natural Selection, a reply to Herbert Spencer", London, 1893.)

Many people are willing to admit that selection explains adaptations, but they maintain that only a part of the phenomena are thus explained, because everything does not depend upon adaptation. They regard adaptation as, so to speak, a special effort on the part of Nature, which she keeps in readiness to meet particularly difficult claims of the external world on organisms. But if we look at the matter more carefully we shall find that adaptations are by no means exceptional, but that they are present everywhere in such enormous numbers, that it would be difficult in regard to any structure whatever, to prove that adaptation had not played a part in its evolution.

How often has the senseless objection been urged against selection that it can create nothing, it can only reject. It is true that it cannot create either the living substance or the variations of it; both must be given. But in rejecting one thing it preserves another, intensifies it, combines it, and in this way creates what is new. Everything in organisms depends on adaptation; that is to say, everything must be admitted through the narrow door of selection, otherwise it can take no part in the building up of the whole. But, it is asked, what of the direct effect of external conditions, temperature, nutrition, climate and the like? Undoubtedly these can give rise to variations, but they too must pass through the door of selection, and if they cannot do this they are rejected, eliminated from the constitution of the species.

It may, perhaps, be objected that such external influences are often of a compelling power, and that every animal must submit to them, and that thus selection has no choice and can neither select nor reject. There may be such cases; let us assume for instance that the effect of the cold of the Arctic regions was to make all the mammals become black; the result would be that they would all be eliminated by selection, and that no mammals would be able to live there at all. But in most cases a certain percentage of animals resists these strong influences, and thus selection secures a foothold on which to work, eliminating the unfavourable variation, and establishing a useful colouring, consistent with what is required for the maintenance of the species.

Everything depends upon adaptation! We have spoken much of adaptation in colouring, in connection with the examples brought into prominence by Darwin, because these are conspicuous, easily verified, and at the same time convincing for the theory of selection. But is it only desert and polar animals whose colouring is determined through adaptation? Or the leaf-butterflies, and the mimetic species, or the terrifying markings, and "warning-colours" and a thousand other kinds of sympathetic colouring? It is, indeed, never the colouring alone which makes up the adaptation; the structure of the animal plays a part, often a very essential part, in the protective disguise, and thus many variations may cooperate towards one common end. And it is to be noted that it is by no means only external parts that are changed; internal parts are always modified at the same time—for instance, the delicate elements of the nervous system on which depend the instinct of the insect to hold its wings, when at rest, in a perfectly definite position, which, in the leaf-butterfly, has the effect of bringing the two pieces on which the marking occurs on the anterior and posterior wing into the same direction, and thus displaying as a whole the fine curve of the midrib on the seeming leaf. But the wing-holding instinct is not regulated in the same way in all leaf-butterflies; even our indigenous species of Vanessa, with their protective ground-colouring, have quite a distinctive way of holding their wings so that the greater part of the anterior wing is covered by the posterior when the butterfly is at rest. But the protective colouring appears on the posterior wing and on the tip of the anterior, to precisely the distance to which it is left uncovered. This occurs, as Standfuss has shown, in different degree in our two most nearly allied species, the uncovered portion being smaller in V. urticae than in V. polychloros. In this case, as in most leaf-butterflies, the holding of the wing was probably the primary character; only after that was thoroughly established did the protective marking develop. In any case, the instinctive manner of holding the wings is associated with the protective colouring, and must remain as it is if the latter is to be effective. How greatly instincts may change, that is to say, may be adapted, is shown by the case of the Noctuid "shark" moth, Xylina vetusta. This form bears a most deceptive resemblance to a piece of rotten wood, and the appearance is greatly increased by the modification of the innate impulse to flight common to so many animals, which has here been transformed into an almost contrary instinct. This moth does not fly away from danger, but "feigns death," that is, it draws antennae, legs and wings close to the body, and remains perfectly motionless. It may be touched, picked up, and thrown down again, and still it does not move. This remarkable instinct must surely have developed simultaneously with the wood-colouring; at all events, both cooperating variations are now present, and prove that both the external and the most minute internal structure have undergone a process of adaptation.

The case is the same with all structural variations of animal parts, which are not absolutely insignificant. When the insects acquired wings they must also have acquired the mechanism with which to move them—the musculature, and the nervous apparatus necessary for its automatic regulation. All instincts depend upon compound reflex mechanisms and are just as indispensable as the parts they have to set in motion, and all may have arisen through processes of selection if the reasons which I have elsewhere given for this view are correct. ("The Evolution Theory", London, 1904, page 144.)

Thus there is no lack of adaptations within the organism, and particularly in its most important and complicated parts, so that we may say that there is no actively functional organ that has not undergone a process of adaptation relative to its function and the requirements of the organism. Not only is every gland structurally adapted, down to the very minutest histological details, to its function, but the function is equally minutely adapted to the needs of the body. Every cell in the mucous lining of the intestine is exactly regulated in its relation to the different nutritive substances, and behaves in quite a different way towards the fats, and towards nitrogenous substances, or peptones.

I have elsewhere called attention to the many adaptations of the whale to the surrounding medium, and have pointed out—what has long been known, but is not universally admitted, even now—that in it a great number of important organs have been transformed in adaptation to the peculiar conditions of aquatic life, although the ancestors of the whale must have lived, like other hair-covered mammals, on land. I cited a number of these transformations—the fish-like form of the body, the hairlessness of the skin, the transformation of the fore-limbs to fins, the disappearance of the hind-limbs and the development of a tail fin, the layer of blubber under the skin, which affords the protection from cold necessary to a warm-blooded animal, the disappearance of the ear-muscles and the auditory passages, the displacement of the external nares to the forehead for the greater security of the breathing-hole during the brief appearance at the surface, and certain remarkable changes in the respiratory and circulatory organs which enable the animal to remain for a long time under water. I might have added many more, for the list of adaptations in the whale to aquatic life is by no means exhausted; they are found in the histological structure and in the minutest combinations in the nervous system. For it is obvious that a tail-fin must be used in quite a different way from a tail, which serves as a fly-brush in hoofed animals, or as an aid to springing in the kangaroo or as a climbing organ; it will require quite different reflex-mechanisms and nerve-combinations in the motor centres.

I used this example in order to show how unnecessary it is to assume a special internal evolutionary power for the phylogenesis of species, for this whole order of whales is, so to speak, made up of adaptations; it deviates in many essential respects from the usual mammalian type, and all the deviations are adaptations to aquatic life. But if precisely the most essential features of the organisation thus depend upon adaptation, what is left for a phyletic force to do, since it is these essential features of the structure it would have to determine? There are few people now who believe in a phyletic evolutionary power, which is not made up of the forces known to us— adaptation and heredity—but the conviction that EVERY part of an organism depends upon adaptation has not yet gained a firm footing. Nevertheless, I must continue to regard this conception as the correct one, as I have long done.

I may be permitted one more example. The feather of a bird is a marvellous structure, and no one will deny that as a whole it depends upon adaptation. But what part of it does not depend upon adaptation? The hollow quill, the shaft with its hard, thin, light cortex, and the spongy substance within it, its square section compared with the round section of the quill, the flat barbs, their short, hooked barbules which, in the flight-feathers, hook into one another with just sufficient firmness to resist the pressure of the air at each wing-beat, the lightness and firmness of the whole apparatus, the elasticity of the vane, and so on. And yet all this belongs to an organ which is only passively functional, and therefore can have nothing to do with the Lamarckian principle. Nor can the feather have arisen through some magical effect of temperature, moisture, electricity, or specific nutrition, and thus selection is again our only anchor of safety.

But—it will be objected—the substance of which the feather consists, this peculiar kind of horny substance, did not first arise through selection in the course of the evolution of the birds, for it formed the covering of the scales of their reptilian ancestors. It is quite true that a similar substance covered the scales of the Reptiles, but why should it not have arisen among them through selection? Or in what other way could it have arisen, since scales are also passively useful parts? It is true that if we are only to call adaptation what has been acquired by the species we happen to be considering, there would remain a great deal that could not be referred to selection; but we are postulating an evolution which has stretched back through aeons, and in the course of which innumerable adaptations took place, which had not merely ephemeral persistence in a genus, a family or a class, but which was continued into whole Phyla of animals, with continual fresh adaptations to the special conditions of each species, family, or class, yet with persistence of the fundamental elements. Thus the feather, once acquired, persisted in all birds, and the vertebral column, once gained by adaptation in the lowest forms, has persisted in all the Vertebrates, from Amphioxus upwards, although with constant readaptation to the conditions of each particular group. Thus everything we can see in animals is adaptation, whether of to-day, or of yesterday, or of ages long gone by; every kind of cell, whether glandular, muscular, nervous, epidermic, or skeletal, is adapted to absolutely definite and specific functions, and every organ which is composed of these different kinds of cells contains them in the proper proportions, and in the particular arrangement which best serves the function of the organ; it is thus adapted to its function.

All parts of the organism are tuned to one another, that is, they are adapted to one another, and in the same way the organism as a whole is adapted to the conditions of its life, and it is so at every stage of its evolution.

But all adaptations can be referred to selection; the only point that remains doubtful is whether they all must be referred to it.

However that may be, whether the Lamarckian principle is a factor that has cooperated with selection in evolution, or whether it is altogether fallacious, the fact remains, that selection is the cause of a great part of the phyletic evolution of organisms on our earth. Those who agree with me in rejecting the Lamarckian principle will regard selection as the only guiding factor in evolution, which creates what is new out of the transmissible variations, by ordering and arranging these, selecting them in relation to their number and size, as the architect does his building-stones so that a particular style must result. ("Variation under Domestication", 1875 II. pages 426, 427.) But the building-stones themselves, the variations, have their basis in the influences which cause variation in those vital units which are handed on from one generation to another, whether, taken together they form the whole organism, as in Bacteria and other low forms of life, or only a germ-substance, as in unicellular and multicellular organisms.*

* The Author and Editor [of the original publication] are indebted to Professor Poulton for kindly assisting in the revision of the proof of this Essay.

Made in the USA
San Bernardino, CA
15 February 2019